Hendrik Bergers

Mediterranean Climate Phenomena

GRIN Verlag

Bibliografische Information der Deutschen Nationalbibliothek:

Die Deutsche Bibliothek verzeichnet diese Publikation in der Deutschen Nationalbibliografie; detaillierte bibliografische Daten sind im Internet über http://dnb.d-nb.de/ abrufbar.

Imprint:

Druck und Bindung: Books on Demand GmbH, Norderstedt Germany
ISBN: 978-3-656-71170-4

This book at GRIN:

http://www.grin.com/en/e-book/277980/mediterranean-climate-phenomena

Ruhr-Universität-Bochum

Fakultät für Geowissenschaften

Geographisches Institut

6.5.2013

Seminar: Regional climates and climate change

SoSe 2013

Term paper:
Mediterranean climate phenomena

Table of contents

1 Introduction

The Mediterranean Sea is an almost enclosed basin with a coverage of 4000 km in west-east and 1200 km in north-south direction.[1] Many European, North African and Middle Eastern states abut on this ocean and are massively influenced by it. The Mediterranean can be denoted as a cutting point between subtropical and humid climate and is influenced by both of them. This and the Mediterranean Sea lead to a very specific and individual climate around this basin.

Figure 1: The Mediterranean Sea

In this paper the leading factors of the climate conditions are revealed, which include references to global atmospheric circulations, regional climate patterns and micro-scale descriptions of specific phenomena in the Mediterranean. Therefore I refer to different scientific articles and mainly to two specific Mediterranean monographs (Branigan; Bolle).

Because of the introducing character of this paper most phenomena are explained in a more general way and do not have reference to direct climate data. More detailed knowledge and data basis can be taken by the references and footnotes.

2 The Mediterranean climate

The Mediterranean climate type is one of the most clearly defined climate zones in the world.[2] It is constituted as the dry-summer subtropical climate, which is discovered around the Mediterranean sea (between latitudes 30° and 45°) and important impacts of the Mediterranean Sea.[3] In the

1 Zeccheto, S.; et al.: Sea surface winds over the Mediterranean basin from satellite data (2000-2004). Meso- and local- scale features on annual and seasonal timescales; in: Journal of Applied Meteorology and Climatology 46: 6 (2007); p. 814-827, here 814.

2 Branigan, J. J.: The Mediterranean Lands; 10th edition; London 1975, p. 31.

3 Lutgens, F.; Tarbuck, E.: The Atmosphere. An Introduction to Meteorology;10th edition, Upper Saddle River 2008, p. 453.

Köppen-classification this is the Csa and Csb (dry-summer subtropical) climate category. It is influenced by subtropical and mid-latitude weather impacts.[4] The variability of Mediterranean climate, which will be illustrated in this paper, is the reason, why geographers often do not use the term "Warm Temperate Western Climate" but the particular term Mediterranean Climate. Hence, there is a massive inner-annual or seasonal variation of climate patterns and a transitional character of this climate zone. Branigan (1975: 31) states 4 essential characteristics of Mediterranean climate: winter rain, summer drought, mild winters and hot summers.

In winter the region is affected by cyclones of the polar front and the Westerlies. Flocas (1984: 255) proves the occurrence of one cold front each weak (compared with one per month in spring, summer and autumn).[5] Moist and warm air masses cominh from the Atlantic reach the Mediterranean and continue eastwards until the Near East (Vb-weather-conditions). So there is high precipitation and mild temperatures. Snow and freezing are rare, because they are effected by local phenomena.

Nevertheless, summer climate is determined by subtropical influence, namely the trade winds, which mean very dry and hot weather periods and aridity. The Azores high causes eastward moving dry air masses, that initiate constant wind streams over the eastern basin (s. Chapter 5).[6] Moreover the Mediterranean is influenced by the Hadley circulation (see chapter 3.1). Both processes` intensity depends not only on the season, but on annual macro-scale atmospheric circulation conditions.[7] Those are the main reasons for hot and dry summers in the Mediterranean.

As a consequence the Mediterranean climate can be matched to both the dry tropics and the humid midlatitudes. Spring and autumn are very brief compared with Central Europe and are influenced by the upcoming season. There can be noticed an increasing gradient of aridity in southern and eastern direction.[8] This is connected with an incline of continental climate phenomena. One recognizes, for example, a great difference between the mild winters in maritime sites like western Iberia and the strong temperature variability in high continental regions like East Anatolia. Moreover the precipitation gradient runs from north-west to south-east. This can be proved by the comparison of the climate station of San Sebastian, where one can find annual humid months, and North Egypt, where <1 month is humid.[9]

---

4 Reddaways, J.; Bigg, G.:Climatic Change over the Mediterranean and links to the more general atmospheric circulation; in: International Journey of Climatology 16 (1996); p. 651-661, here 651. Lutgens, Tarbuck 2008: 454.

5 Flocas, A. A.: The annual and seasonal distribution of fronts over Central Southern Europe and the Mediterranean; in: Journal of Climatology 4: 3 (1984); p. 255-267, here 255.

6 Wagner, H. G.: Mittelmeeraum; Darmstadt; 2., vollständig überarbeitete Auflage 2011, p. 119.

7 Bolle, H. J.: Mediterranean Climate. Variability and trends; Berlin 2003, p. 25.

8 Borman, M.: Das Mittelmeerklima. Erscheinungen und Ursachen; in: Geographie heute 52 (1987), p. 11-23, here 11.

9 Wagner 2011: 119-120.

2.1 Temperature

For the temperature, there must be underlined a specific spatial distribution in the Mediterranean region. In the Northern part the land surface temperature is on average about 5° C lower than in the Southern areas. In cold mountain areas of the Alps one can find annual mean temperatures of -2° degree whereas in the hot lowlands of North Africa annual average temperatures of 22° degree occur. In the more coastal region the warmest months are delayed from July to August because of the balancing maritime effect. Moreover, the northern part of the Mediterranean must be specified. The central and eastern regions occur warmer than the western ones or the Aegean gulf. Reasons for this are the lower northern radiation, rising cool water streams caused by internal effects and the income of cooler water through rivers and the Black Sea.[10]

2.2 Precipitation

The Mediterranean is constituted by a large-scale decreasing intensity of precipitation northwesterly to southeasterly. The dryness of southeast Mediterranean regions is connected with a much more intensive total evaporation compared with northwest patterns. Hence, the precipitation gradient is modified by meso- and micro-scale impacts, which is in the Mediterranean mainly the relief (almost two thirds are orographic rain causing mountains). So there is a great local diversity of precipitation levels.[11] In some areas (Anatolia or Algeria) a continental climate, which is caused by orographic rain shadows, can be found.[12] Xoplaki et al. (2004: 63) disclaim a connection of Mediterranean sea surface temperature and precipitation variability patterns. They carve out the connection between the North Atlantic Oscillation (NAO) and local precipitation amounts.[13]

Here a short explanation about the NAO is given. The NAO refers to the north southern distribution of air masses (pressure gradient) between two centers of action: the Icelandic Low and the Azores High. This phenomenon has its greatest occurrence in winter, when the atmosphere is dynamic. In this season the NAO is responsible for more of one third of the North Atlantic pressure variances. Hurrell[14] and Wanner[15] et al. carried out, that the NAO has massive impacts on the temperature patterns and the precipitation and storm amounts of the Northern Hemisphere. Since Flohn et al. (1998: 23) a strong NAO index leads to a warming and a raising moisture in Southern

10 Bolle 2003: 34.
11 Wagner 2011: 120.
12 Branigan 1975:32.
13 Xoplaki, E.; et al.: Wet season Mediterranean precipitation variability. Influence of large scale dynamics and trends; in: Climate Dynamics 23: 1 (2004); p. 63-78, here 63.
14 Hurrell, J.W.: North Atlantic Oscillation (NAO); in: Encyclopedia of Ocean Science 4 (2001); p. 1904-1911.
15 Wanner, H.: North Atlantic Oscillation. Concept and Studies; in: Surveys in Geophysics 22 (2001); p. 321-382.

Europe and the Mediterranean.[16]

Another phenomenon, which is connected to Mediterranean precipitation patterns, is El Nino Southern Oscillation (ENSO), which describes the properties of the Pacific atmosphere ocean circulation, but has large macro-scale impacts. So precipitation anomalies in Southern Europe can be assigned to different stages of ENSO. Most effects of the ENSO can be recognized in spring and autumn.[17]

Moreover it has to be mentioned, that the rainy periods differ southwards. In South Europe a longer precipitation stage can be measured. Contradictory to this is South Mediterranean precipitation, which is cumulative concentrated on short and intensive periods. Here lies an erosional risk, because water is rarely infiltrating and flowing on the marginal planted soil. Another precipitation phenomenon in North Africa is the occurrence of a couple of dry years in a row.[18]

3 Seasonal variability

The most obvious phenomenon of Mediterranean climate is a great seasonal variability. This is caused by the affects of two global circulation systems, namely the westerlies in the winter and the Inner Tropical Convergence Zone (ITCZ) in the summer. The seasonal climate occurrences in meso- and macro-scales are presented in the following chapter.

3.1 Summer conditions

In the summer, the Mediterranean is influenced by the Hadley circulation, which is determined by heated, equatorial, moist air, which moves polewards. Therefore, the Mediterranean is affected by maritime tropical (mT) and continental tropical (cT) air masses.[19] In the space of the ITCZ, which is moving northwards in the summer, the moist air rains and sinks down at the Tropic of Cancer. The intensity of this process is determined by other, global atmospheric processes, which could be the Southern Oscillation or the Indian Monsoon, but these connections should not be discussed here. In wintertime he westerlies are further north and do not affect the Mediterranean climate intensive, but just in some rare "blocking" situations.[20]

Northerly winds are determining the summertime climate conditions. In the western basin this is

16 Flohn, H.: Behavior of the Centers of action above the Atlantic since 1881. Part II. Associations with regional climate anomalies; in: International Journal of Climatology 18: 1 (1998), p. 23-36, here 23.

17 Zeng, N.; Lau, K. M.: Euro-Mediterranean rainfall and ENSO. A seasonally varying relationship; in: Geophysical Research Letters 21: 12 (2002), p. 59.1-59.4, here 59.1, 59.4.

18 Wagner 2011: 120.

19 Branigan 1975: 34.

20 Bolle 2003: 27.

caused by the Azores High, which is moving northwards in spring, and north or north-easterly trade winds are generated. The eastern basin is influenced by the Asiatic low pressure monsoonal circulation, which means north or north-westerly winds.[21]

The inner Mediterranean dynamics are caused by the temperature differences between the ocean and the surrounding land mass with a specific relief. In interaction with the large scale effects the ocean continent distribution is important. In this context the high ratio of mountain regions is fundamental, because lots of local wind and precipitation systems, which are of course determined as well by general circulation systems of the North Atlantic and the Mediterranean, are excited from this. Some semi-consistent high pressure systems constrict moist warm air to reach the condensation level, which causes drought.

Soils, which are less vegetated because of human impacts or water shortage, often develop to heat lows in the intensive radiation in summer. In spite of the generated convection the summer sky is mostly clear without any clouds. Bolle (2003: 28-29) gives two reasons for this. The first reason is, that the air warms up very fast, the relative moisture is decreasing and even wet maritime air becomes drier. The other reason is found in an overlapping high pressure system, which prevents the air to rise and condensate. As a consequence the coastal zones, where agriculture is accumulated, are dominated by dry convection. Some special situations, meteorological caused by wooden hinterlands, can lead to summertime thunderstorms.[22]

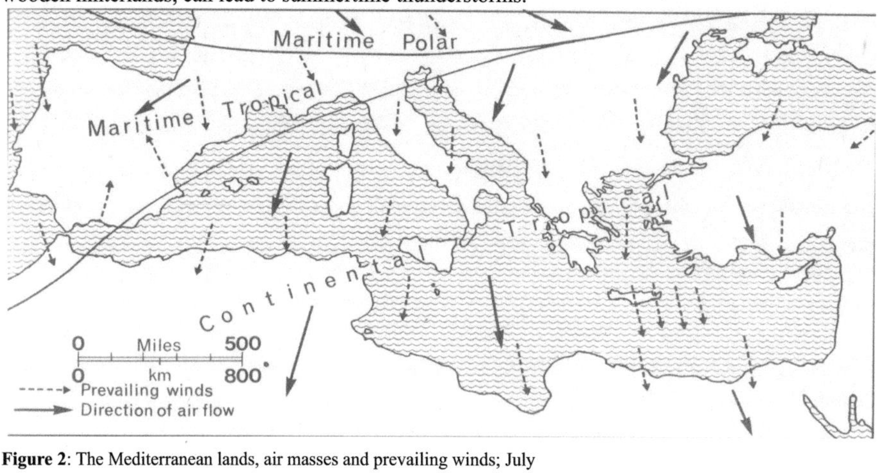

Figure 2: The Mediterranean lands, air masses and prevailing winds; July

In spring and summer time the land-sea-circulation reveals much impact on climate conditions in

21 Branigan 1975: 36.
22 Bolle 2003: 28-29.

the Mediterranean, because it is mostly responsible for the interchange of humid and arid air masses. Strong thermal differences are growing between land and sea, which initiates warm, coastal air to rise and generate low pressure at the land. Because of this the maritime air flows landwards and warms up rapidly. So the relative moisture is decreasing and more effects are necessary to observe rain. This could be, for example, an orographic rise or a mixture with dry convection to move upwards until the condensation level. In local different systems, for example the Apennines, the land-sea-circulation often develops to cumulus convection.[23]

3.2 Winter conditions

In the winter the Mediterranean climate is strongly changing and effected by complete different effects than in the summer. As the ITCZ is moving southwards, maritime Polar (mP) air masses influence the Mediterranean.[24]

Between the two centers of action, which are known as the Icelandic Low and the Azores High, a jet-stream occurs. These are the westerlies. Rossby waves (caused by orographic barriers) affect the westerlies' influence on the Mediterranean. These waves are caused by "semi-permanent pressure systems".[25] Because of the pressure circulation influenced by specific pressure cells the atmospheric flow moves eastwards. The North Atlantic pressure system, which is public as the North Atlantic Oscillation, is moreover determined by different oceanic circulations of cold and warm water flows. [26]

The westerlies' low pressure cells often travel the whole Mediterranean length from west to east and bring moist air masses.[27] Another reason for the pronounced wintertime depression over the Mediterranean Sea is the sea surface temperature, which is on average about 2° degree warmer than comparable water in this latitude. So the relative warm air above the ocean is a constant pool of low pressure and 73% of the Mediterranean depressions origin above the ocean, mainly the western basin.[28]

A relevant wintertime phenomenon of the Mediterranean are extra-tropic cyclones, which are caused by the high frequently occurrence of depressions and dominate the Mediterranean climate conditions in this season. The main areas of intensive cyclogenesis and cyclone reinforcement are

23 Bolle 2003: 30.
24 Branigan 1975: 36.
25 Bolle 2003: 25.
26 Bolle 2003: 31.
27 Branigan 1975: 31.
28 Bethoux, J. P.; et al.: The water circulation in the North Western Mediterranean Sea, its relations with wind and atmospheric pressure; in: Nihoul, J. C. J. (editor): Hydrodynamics of semi-enclosed seas; Amsterdam 1982, p. 129-142, here 138, see also Branigan 1975: 37.

the Gulf of Genoa, the area south of the Atlas Mountains (North West Africa) and the Middle East. Another important area of cyclogenesis is, of course, the Mediterranean Sea itself. As already mentioned, the Mediterranean Sea is in wintertime relative warm and generates depression cells. The western cyclogenesis areas seem to occur in connection with a lee condition. The Middle East cyclones are determined by the Asian monsoon.[29] Compared with other Northern Hemisphere cyclones the Mediterranean ones do not have a large spatial distribution and short life cycles, but in the almost enclosed Mediterranean the play an important role.[30]

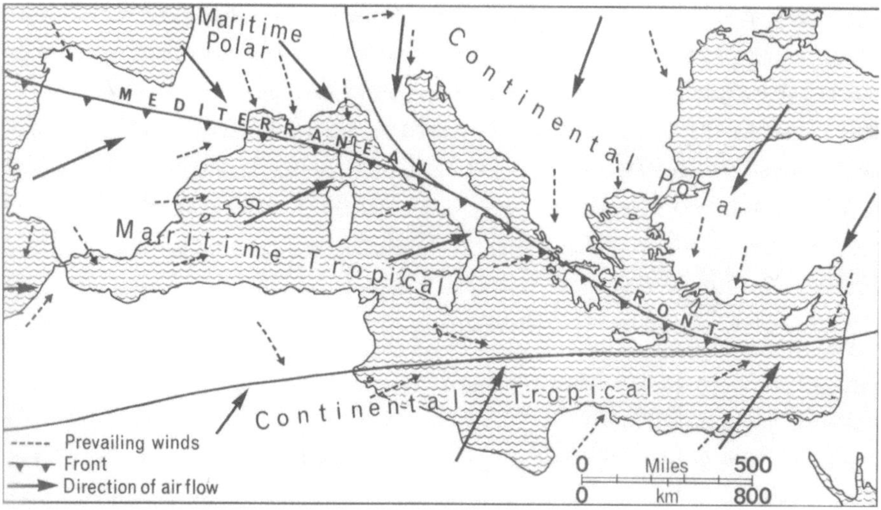

Figure 3: The Mediterranean lands, air masses and prevailing winds; January

In the autumn the regional land-sea-circulation turns around and because of the special heat of water the ocean is warmer than the continental regions. So the convection often starts above the ocean and can lead with strong latent heat fluxes to thunderstorms, which move landwards. In this season the contrast between plain and mountain surface temperature can cause locally specific air circulation between the lowlands and rock masses.[31]

4 The impact of the Mediterranean Sea

Compared with other Mediterranean climate regions (South Africa, Australia) the European-

29 Trigo, I. F.; Tavies, T. D.: Objective climatology of cyclones in the Mediterranean region; in: Journal of Climate 12: 6 (1999); p. 1685-1696, here 1685, 1695. See also Trigo, I. F.; et al.: Climatology of cyclogenesis mechanisms in the Mediterranean; in: Monthly Weather Review 130: 3 (2002), p. 549-569, here 459.

30 Trigo 1999: 1695.

31 Bolle 2003: 31.

African-Asian Mediterranean climate is not only constricted on small coast-belts, but very expensive, which can be lead to the massive influence of the Mediterranean Sea. There is the fact, that the surrounding land masses have in contradictory to the sea a negative radiation flux balance. Hence, the positive radiation budget of the ocean is important for the land masses around, which gain an influx of heat to adjust the negative radiation budget. Many coastal areas are energized by land-sea-circulation systems, that release energy by condensation processes. Especially in the east, more dry basin the required water for the vegetation comes from the ocean and is often transported in hinterland areas by land-sea-circulation. Furthermore the maritime winds cool down the coastal regions, because the summer daytime air from the ocean is cooler than the continental one and the warming air increases evaporation because of the low relative moisture.[32]

Here I want to give a short excursion to the so called "Mediterranean conveyor belt" (termohaline circulation), which means the water flows in the ocean. Through the Strait of Gibraltar there is an intrusion of warm and fresh Atlantic waters, which are moved eastwards by the westerlies. A transformation to salty and cold water, which is heavy and moves westwards in a depth of 500 m, occurs.[33] Since Bethoux et al. (1982: 129) these ocean flows are affected by internal effects and pressure or wind systems do not have much impact.[34]

5 Mediterranean winds

Generally there must said, that the eastern and the western Mediterranean basin are influenced by different processes, which can be shown by the number of fronts, for example.[35] In this context the spatial diversity and local complexity of meso- and micro-scale processes and effects, mainly induced by coastal mountains, is underlined. In fact of this Zeccheto (2007: 814) diagnoses a lack of studies on this.[36]

In the eastern basin, mainly in the Aegean, the so called Meltemi wind often occurs in summertime. The Meltemi (or Etesien) is a northeastern to northwestern, very dry and cold (continental) wind, which is often stormy above the ocean. The genesis of this wind origins in the summer high pressure zones in Italy or Balkan (see figure 4), but often high pressures between Germany and Russia lead to this phenomenon, too. In Turkey and Greece are due to the heat mainly low pressure pools found, which lead to a west-east pressure gradient and air mass movement.[37]

---

32 Bolle 2003: 33-34.
33 Bolle 2003: 35.
34 Bethoux et al. 1982: 129.
35 Reddaway 1996: 651.
36 Zeccheto 2007: 814.
37 Walter, P.O.: Meltem. Ein griechisch-türkischer Sommerwind. Erläuterungen zu seiner Entstehung; 2004. <http://www.esys.org/wetter/meltemi.html> [2.5.2013; 21:53]

Another fundamental wind in the Mediterranean is the Bora, which occurs in the Adriatic and flows from northeast. His characteristics can be compared to Meltemi, because Bora is cold and dry, too. A difference that should be marked out are the enormous wind speeds of Bora. Moreover it is connected with Central European high pressure centers,. Offshore cyclones can cause Bora winds, when they draw the dry continental air above the coastal mountains.[38]

Figure 4: Genesis of the Meltemi

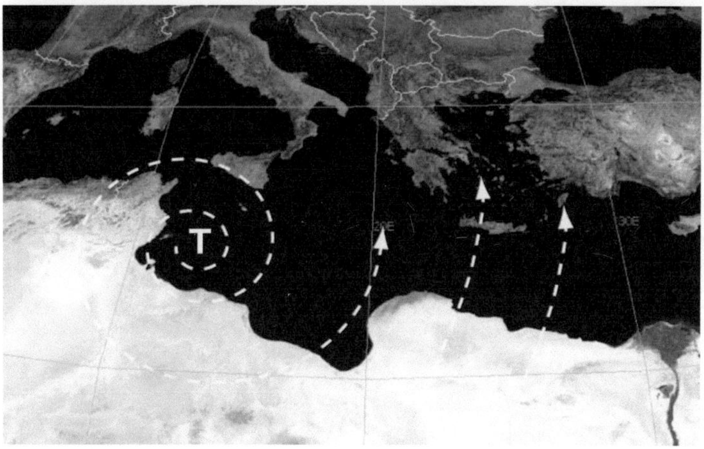

Figure 5: Genesis of the Scirocco

Finally the Scirocco wind is presented. As it is shown on figure 5 the Scirocco has its source in the

38 Grisogono, B.; Belusic, D.: A review on recent advances the meso- and microscale properties of the severe Bora wind; in: Tellus A 61: 1 (2009); p. 1-16, here 1-2.

North African Sahara, where low pressure zones origins because of the enormous heat. So the contra clockwise rotating of those depressions and the high pressure zones in the North Mediterranean force air to move northwards. Because of its Saharan origin the characteristics of this wind are trivial: Scirocco brings hot and dry air, which can include desert sand. Above the ocean the Scirocco often becomes wetter.[39]

The represented winds are only a selection of a large number of different scaled winds and atmospheric circulations that occur in the Mediterranean. It is shown that most winds are determined by semi-permanent pressure gradients, which lead to a relative continuity, but seasonal variability, of air movement in the Mediterranean.

6 Conclusion

In this paper the specific phenomena of the Mediterranean are depicted. The Mediterranean is a very complex and variable region that is influenced by many different effects and causes. The most important fact is the seasonal variability of the influence of global circulation systems. This means the determining impact of the westerlies in the winter and the ITCZ in the summer. As it is worked out here the effective climate conditions can be matched to those phenomena. Moreover a capillary spatial distribution of climate patterns is worked out. Those effects are caused by relief, atmospheric or surface impacts.

39 Walter, P. O.: Winde im Mittelmeer. Der Scirocco; 2004. <http://www.esys.org/wetter/shirokko.html > [3.5.2013; 17:10]

7 References

7.1 Literature

Bethoux, J. P.; et al.: The water circulation in the North Western Mediterranean Sea, its relations with wind and atmospheric pressure; in: Nihoul, J. C. J. (editor): Hydrodynamics of semi-enclosed seas; Amsterdam 1982, p. 129-142.

Bolle, H. J.: Mediterranean Climate. Variability and trends; Berlin 2003.

Borman, M.: Das Mittelmeerklima. Erscheinungen und Ursachen; in: Geographie heute 52 (1987), p. 11-23.

Branigan, J. J.: The Mediterranean Lands; 10th edition; London 1975.

Flocas, A. A.: The annual and seasonal distribution of fronts over Central Southern Europe and the Mediterranean; in: Journal of Climatology 4: 3 (1984); p. 255-267.

Flohn, H.: Behavior of the Centers of action above the Atlantic since 1881. Part II. Associations with regional climate anomalies; in: International Journal of Climatology 18: 1 (1998), p. 23-36.

Grisogono, B.; Belusic, D.: A review on recent advances the meso- and microscale properties of the severe Bora wind; in: Tellus A 61: 1 (2009); p. 1-16.

Hurrell, J.W.: North Atlantic Oscillation (NAO); in: Encyclopedia of Ocean Science 4 (2001); p. 1904-1911.

Lutgens, F.; Tarbuck, E.: The Atmosphere. An Introduction to Meteorology;10th edition, Upper Saddle River 2008.

Reddaways, J.; Bigg, G.: Climatic Change over the Mediterranean and links to the more general atmospheric circulation; in: International Journey of Climatology 16 (1996); p. 651-661.

Trigo, I. F.; Tavies, T. D.: Objective climatology of cyclones in the Mediterranean region; in: Journal of Climate 12: 6 (1999); p. 1685-1696.

Trigo, I. F.; et al.: Climatology of cyclogenesis mechanisms in the Mediterranean; in: Monthly Weather Review 130: 3 (2002), p. 549-569.

Wagner, H. G.: Mittelmeeraum; Darmstadt; 2., vollständig überarbeitete Auflage 2011, p. 119.

Wanner, H.: North Atlantic Oscillation. Concept and Studies; in: Surveys in Geophysics 22 (2001); p. 321-382.

Xoplaki, E., et al.: Wet season Mediterranean precipitation variability. Influence of large scale dynamics and trends; in: Climate Dynamics 23: 1 (2004); p. 63-78.

Zeccheto, S.; et al.: Sea surface winds over the Mediterranean basin from satellite data (2000-2004). Meso- and local- scale features on annual and seasonal timescales; in: Journal of Applied

Meteorology and Climatology 46: 6 (2007); p. 814-827.

Zeng, N.; Lau, K. M.: Euro-Mediterranean rainfall and ENSO. A seasonally varying relationship; in: Geophysical Research Letters 21: 12 (2002), p. 59.1-59.4.

7.2 Internet sources

Walter, P.O.: Meltem. Ein griechisch-türkischer Sommerwind. Erläuterungen zu seiner Entstehung; 2004. <http://www.esys.org/wetter/meltemi.html> [2.5.2013; 21:53]

Walter, P. O.: Winde im Mittelmeer. Der Scirocco; 2004. <http://www.esys.org/wetter/shirokko.html > [3.5.2013; 17:10]

7.3 Figures

Figure 1: <http://commons.wikimedia.org/wiki/File:Mediterranean_Sea_location_map.svg > [3.5.2013; 19:45]

Figure 2: Branigan, J. J.: The Mediterranean Lands; 10th edition; London 1975, p. 35.

Figure 3: see Figure 2

Figure 4: <http://www.esys.org/wetter/meltemi-nasasat-hq.jpg> [3.5.2013; 19:45]

Figure 5: <http://www.esys.org/wetter/shirokko-map.jpg> [3.5.2013; 19:45]